"FLEXOGRAPHY 101 Booklets" are published to disseminate information on topics of interest to flexographers. The contents of this booklet are a derivative of FLEXOGRAPHY: Principles and Practices 6.0 and all respective credit is due to its authors.

FTA Technical Education Services Team
3920 Veterans Memorial Hwy Ste 9, Bohemia, NY 11716-1074

INTRODUCTION

Color has been defined as "the perception of light that has been modified by an object." This definition actually refers to more than color. It alludes to what determines color, a light source, an object and observer (Image 1).

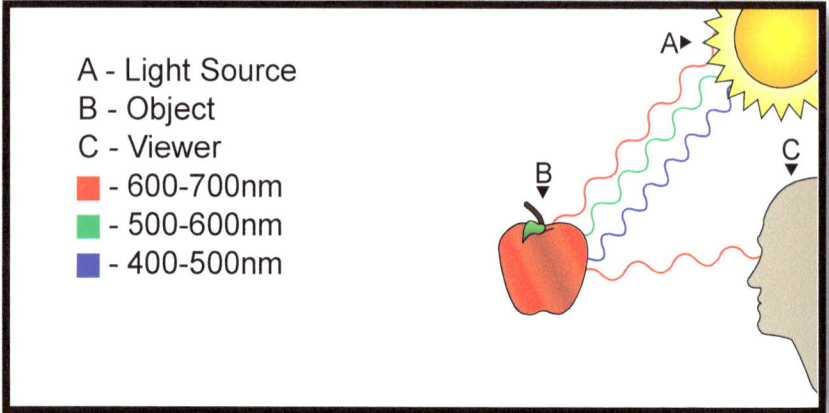

A - Light Source
B - Object
C - Viewer
■ - 600-700nm
■ - 500-600nm
■ - 400-500nm

Image 1: Light, Object Observer

Note: Light comes from a source and is modified not only by the object being observed, but also by the surroundings.

The first element to examine is light itself. The light we see is part of a natural phenomenon that includes x-rays, ultraviolet radiation, visible light, infrared radiation, television and radio waves. The key word is waves. All are a class of what is called electromagnetic radiation and the key difference is in the wavelength. X-rays have the shortest wavelength and radio the longest.

Visible light ranges in wavelength from approximately 400 to 700 nanometers (nm). White light contains an equal amount of all of these wavelengths. It can be broken out or dispersed, such as with a glass prism, into light of the separate wavelengths that make up the "colors of the rainbow" (Image 2). All visible light is a combination of these wavelengths.

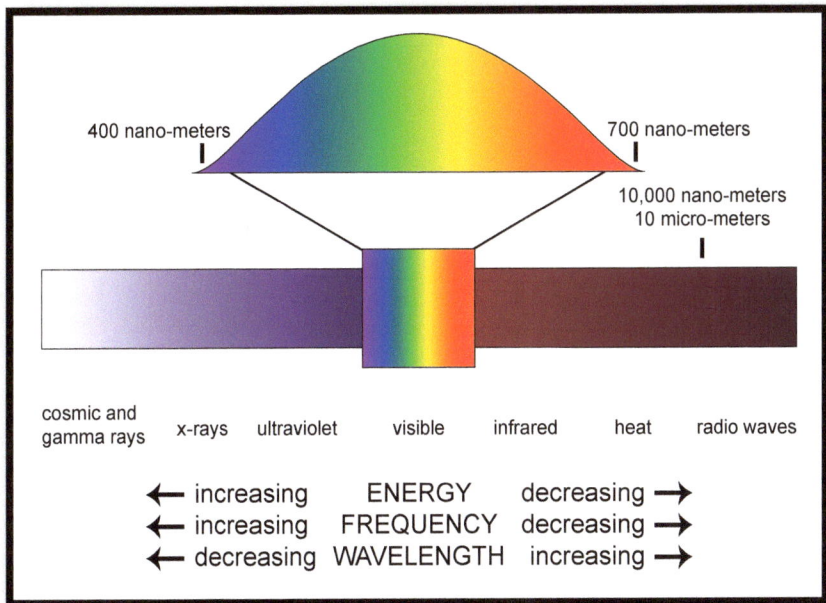

Image 2: Eletromagnetic Spectrum with Visible Spectrum Showcased

HUMAN COLOR VISION

The spectrum is divided into the red, green, and blue regions because this matches the way the human eye sees color. The eye has three sensors or receptors that detect the three primary colors. All colors perceived are a mixture of these primary colors. The spectral-response functions of the eye are shown in (Image 3). They are based on experiments conducted by the CIE and represent the standard observer. Due to considerations such as age, culture, the rods and cones in our eyes, each person will see color differently. Therefore, the perception of color represents an uncontrollable variable.

If an object is viewed using other than a white-light source, the perceived color will take on the hue of the illumination. While sunlight is the most natural way to view objects, it is not an ideal light source to judge the color of objects; it is simply too variable. Artificial light sources are available and may be controlled and

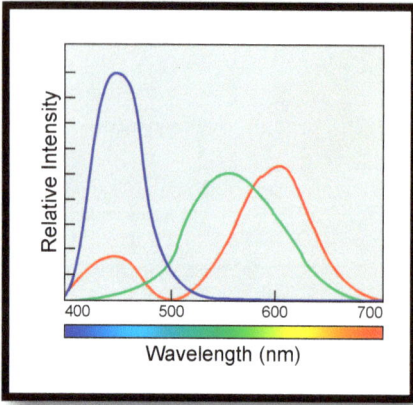

Image 3: Spectral-response functions of the eye

specified to simulate average, natural daylight, and incandescent and fluorescent lamps. A scale of color temperature, expressed as degrees Kelvin (°K), is used to quantify light sources. Various artificial light sources have color temperatures that range from about 4,000° K to 6,800° K. The D50 CIE Standard Illuminant has a color temperature of 5,000° K, representing average natural daylight, and is the color temperature most widely used in graphic arts viewing booths. Image 4 shows three light sources along with their spectral curves. The spectral curve shows the amount of light of the source throughout the visible range of wavelengths, which is roughly from 400 to 700 nanometers.

Light Source, Object, Observer

Color results from the interaction between light source, an object and the observer (Image 1). The human observer sees this modified light and perceives it as a distinct color. All three elements, light, object and observer, must be present for color, as we know it to exist. If an object is viewed using other than a white-light source, the perceived color will take on the hue of the illumination.

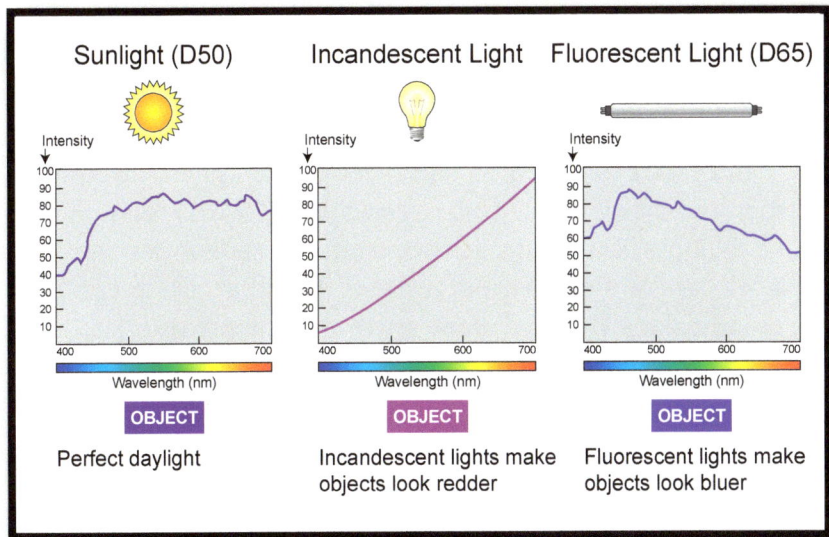

Sunlight (D50)	Incandescent Light	Fluorescent Light (D65)

Perfect daylight

Incandescent lights make objects look redder

Fluorescent lights make objects look bluer

Image 4: Light Sources

Light is the Origin

Color perception is a result of achromatic and chromatic signals derived from light stimuli.

- Chromatic signals (~ hue) are mostly derived from the light's wavelength.
- Achromatic signals (~ lightness / grayscale vision) are mostly derived from the light's energy at a certain wavelength.

The eye has photoreceptor cells, called cones, which we divide into three types. S-Cones are sensitive to short-wavelength light ("blue"). M-Cones are sensitive to medium-wavelength light ("green"). L-Cones are sensitive to long-wavelength light ("red"). Each type of cone is more or less stimulated by a light beam and passes on its stimulus value.

The next step is to transform these values into values of three opponent-colors systems.

The luma system (~ range from black to white, or dark to fully illuminated, respectively), the blue/yellow system and the red/green system.

Finally, the values of the three opponent-colors systems are put together and make up our color perception. Of course, we absorb multiple light beams at once from different angles and therefore can form an image of a scene. The grayscale images on the right show (Image 5) how much visual information is contained in each channel.

An interesting aspect of cones is their different sensitivity in regards to wavelength. Image 5 shows the sensitivity curves for each type of cone.

We see that cones react to a wide range of wavelength and how their curves overlap. We can also see that sensitivity-peaks are not evenly spaced. This demonstrates that for most wavelengths we need the input of all three types of cones to distinguish changes in wavelength and/or intensity (number of photons). Furthermore, Image 5 provides evidence that the medium and long-wavelength cones contribute the most to our brightness perception, if we consider the formula of the luma system from the above: $L = B + (R + G)$.

The physiological prerequisites allow drawing the conclusion that there is no uniformity in color perception among individuals. Several physiological deviations can cause a shift in color perception. For individuals with "normal" vision these shifts can be hardly measured. Less fortunate individuals suffer from significant shifts in color perception, called color vision deficiency or "color blindness". Most of the times this is caused by cones which are defect, absent, or altered in their spectral sensitivity.

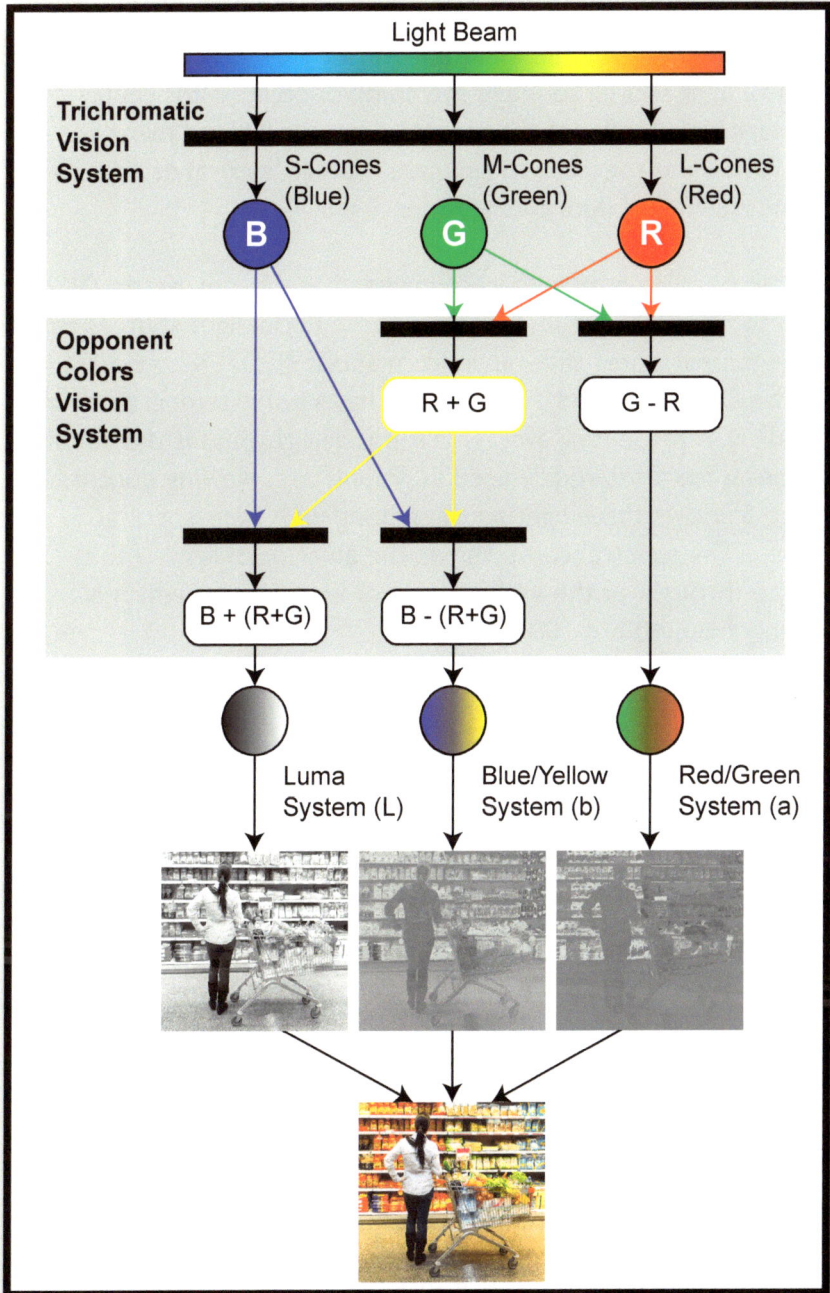

Image 5: Example of How the Human Eye Sees Color (Rods and Cones)

Light Source

While sunlight is the most natural way to view objects, it is not an ideal light source to judge the color of objects; it is simply too variable. Artificial light sources are available and may be controlled and specified to simulate average, natural daylight, and incandescent and fluorescent lamps.

A scale of color temperature, expressed as degrees Kelvin (°K), is used to quantify light sources. Various artificial light sources have color temperatures that range from about 4,000° K to 6,800° K. The D50 CIE Standard Illuminant has a color temperature of 5,000° K, representing average natural daylight, and is the color temperature most widely used in graphic arts viewing booths. Image 5 shows three light sources along with their spectral curves. The spectral curve shows the amount of light of the source throughout the visible range of wavelengths, which is roughly from 400 to 700 nanometers.

Surroundings

A surface or surrounding does not have a precise color, but rather an ability to absorb certain wavelengths and reflect others. A leaf appears green because it absorbs red and blue and reflects only green. It always appears green as long as the light source doesn't change. The reflection of light from a colored surface has two components. Some light is spectrally reflected from the first layer of the surface or surrounding.

The remainder enters the substrate and undergoes scattering and multiple reflections before reemerging from the material as a diffuse reflection. When the light meets a pigment particle some wavelengths are absorbed, while others are reflected. The emerging light is perceived to have a color corresponding to the unabsorbed wavelengths. Since the light reflected from the first layer has not been absorbed, it makes the surface appear lighter.

Color Deficiency

Color vision deficiency is the inability to distinguish certain shades of color or in more severe cases, see colors at all. The term "color blindness" is also used to describe this visual condition, but very few people are completely colorblind.

Red-green deficiency results in the inability to distinguish certain shades of red and green. Most people with color vision deficiency can see colors, but they have difficulty differentiating between particular shades of reds and greens (most common) or blues and yellows (less common).

People who are totally color blind, a condition called achromatopsia, can only see things as black and white or in shades of gray. The severity of color vision deficiency can range from mild to severe depending on the cause. It will affect both eyes if it is inherited and usually just one if the cause for the deficiency is injury or illness. Color vision is possible due to photoreceptors in the retina of the eye known as cones. These cones have light sensitive pigments that enable us to recognize color. Found in the macula, the central portion of the retina, each cone is sensitive to either red, green or blue light, which the cones recognize based upon light wavelengths.

Normally, the pigments inside the cones register differing colors and send that information through the optic nerve to the brain enabling you to distinguish countless shades of color. But if the cones lack one or more light sensitive pigments, you will be unable to see one or more of the three primary colors thereby causing a deficiency in your color perception.

The most common form of color deficiency is red-green. This does not mean that people with this deficiency cannot see these colors at all; they simply have a harder time differentiating between them. The difficulty they have in correctly identifying them depends on how dark or light the colors are.

Another form of color deficiency is blue-yellow. This is a rare and more severe form of color vision loss than red-green since persons with blue-yellow deficiency frequently have red-green blindness too. In both cases, it is common for people with color vision deficiency to see neutral or gray areas where a particular color should appear.

COLOR MODELS

A color model is an abstract mathematical model describing the way colors can be represented as tuples of numbers, typically as three or four values or color components. When this model is associated with a precise description of how the components are to be interpreted (viewing conditions, etc.), the resulting set of colors is called color space.

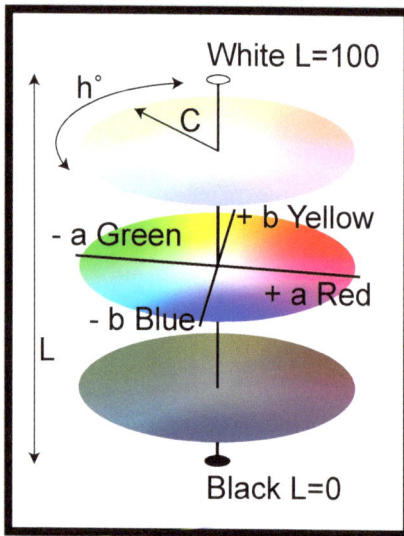

Image 6: CIE Perceptual Color Space

CIE Color Space

Each of the three components of color (source, object, observer) has a specific spectral response curve. These combine to give the final response curve. Rather than specifying color in terms of this final spectral curve, it is more useful to combine them mathematically and create a three-dimensional color space called the CIE perceptual color space (Image 6). In this space, three numbers uniquely specify a color. The system is not perfect however; a unique color in CIE perceptual color space can be formed by more than one combination of source, object, and observer.

CIE perceptual color space is the basis of quantitative color. There are different mathematical algorithms for combining the spectra leading to different numbers, but all have the general appearance of the model shown in Image 6. In 1976, the CIE (Commission Internationale de l'Eclairage) standardized the model called L*a*b* and the model is commonly referred to as the CIEL*a*b* color space. The additional descriptive term "perceptual" means that this color space is based on how the eye perceives color. This is in contrast to name or ingredient based description of color such as "warm red" or 0 percent Cyan, 100 percent Magenta, 90 percent Yellow, and 0 percent Black as seen in many descriptions of named colors.

Every color an observer can see can be represented by its location in CIE perceptual color space, which is commonly described as L*a*b* and L*C*h°.

L*a*b*
L stands for lightness and is the vertical dimension in color space. Every color has a lightness or L value. Unlike L, a and b do not stand for a perceptual attribute. Instead, they are the x, y coordinates of the chromatic plane. The chromatic plane is a cross section of perceptual space viewed from the top as a two-dimensional plane. Every color has a location in this plane (Image 7). The red color indicated by the circle is located 75 units in the "a" direction and 33 units in the "b" direction. The easiest way to envision this plane is to think about the face of a clock. 3:00 o'clock is 0 degrees. 12:00 o'clock is 90 degrees, 9:00 o'clock is 180 degrees, and 6:00 o'clock is 270 degrees.

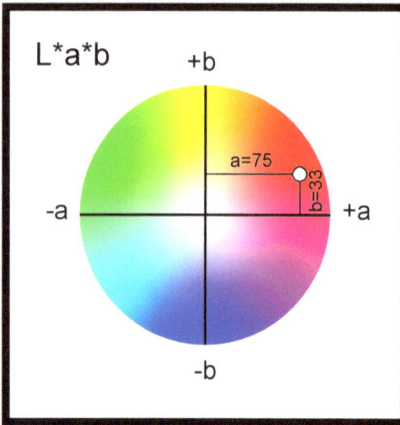

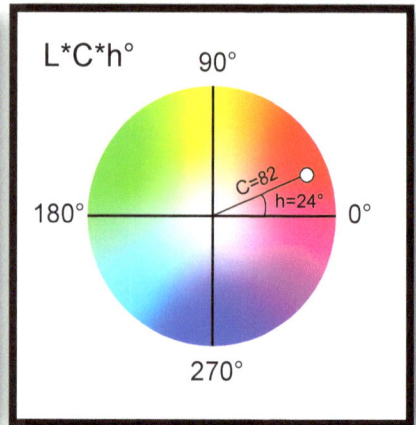

Image 7: L*a*b* Chromatic Plane   Image: 8: L*C*h° Chromatic Plane

L*C*h°

Referring to the same red color as in Image 7, L*C*h° is simply a different way of describing and navigating to that color as shown in Image 8. This time, however, the color is reached by going out 82 units (c) at a 24-degree angle (h). Geometrically, the difference between L*a*b* and L*C*h° is the difference between a cartesian and polar coordinate system. The much more important difference is that L*C*h° represents the perceptual attributes of color. These attributes are described as follows:

L, or lightness, is the lightness or darkness of the color. The scale goes from 0 for black to 100 for white.

C, or chroma, refers to the saturation of the color; zero along the central vertical axis. A color with a C of 0 is neutral or gray. The more saturated or pure the color, the higher the C value. Another descriptive word used is a strong color as opposed to a weak color. Values are not capped at any particular value but rarely exceed 100.

h, or hue, is the perception of the "color" attribute of color. This may seem like a circular definition but the best way to describe hue is to say it determines whether the color is red or green or purple. The hues are arranged in a circular fashion so that a particular direction represents a specific hue.

Gamut Mapping

The range of colors that can be reproduced by C, M, and Y inks on a particular substrate is called the gamut of the system. Different combinations of the process colors are used to create all printing colors. Even if inks of the "perfect" CMY shown were available, one still could not combine them to create all colors that the human eye can perceive. A simple and real life example would be a red laser, the kind used in the supermarket to scan the bar codes. This has a light of only one wavelength. The spectrum is a sharp spike at 633 nanometers. The ultrapure red color of the laser beam is considered an out-of-gamut color, and there is no way even "perfect" C, M, and Y inks could be combined to yield such a spectrum.

Every device, including monitors, scanners and proofers has their own specific gamut; colors they can read or render. Image 9 shows the gamut of a digital proofing system and a flexo press. As is the case in this example, the proofing system usually has a larger gamut than the press.

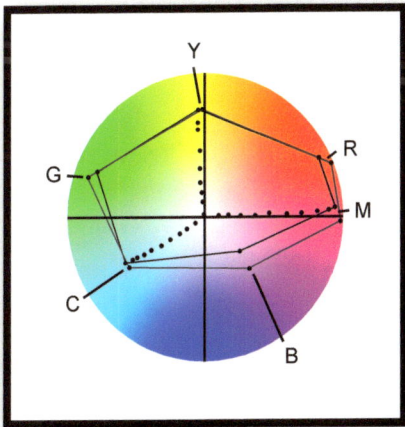

Image 9: Multiple Gamuts Mapped on the L*C*h° Chromatic Plane

This is particularly true for high-end proofing systems used to make the final contract proof. In Image 9, Y, M, C are the points of 100 percent yellow, magenta and cyan. R, G, B, are the solid overprints of YM, CM and YC respectively. The colors, which are inside the polygon connecting these six points make up the gamut of the device. On this same illustration (Image 9) the gamut of a color transparency would be larger still.

The dots inside the diagram in Image 9 the C* (chroma) and h° (hue) locations for colors produced with different dot percentages of the process-color inks. The points are actual measurements of the L*C*h° values and clearly demonstrate that the hue remains constant when printing different tones of the same color. When printing on white paper, only the chroma and lightness should change. Image 9 also illustrates that this indeed happens in the real world.

Gamut mismatch is one of the great challenges in printing, and boils down to the question of what do to with the colors that are outside the gamut of the printing press. When reproducing a color transparency, for example, there are many colors the press cannot reproduce. The gamut of the transparency must be compressed. The methods to do this are what much of color management, scanner setup, and the conversion of RGB to CMYK is all about.

ELECTROMAGNETIC SPECTRUM
Electromagnetic waves are waves that are capable of traveling through a vacuum. Unlike mechanical waves that require a medium in order to transport their energy, electromagnetic waves are capable of transporting energy through the vacuum of outer space. Electromagnetic waves are produced by a vibrating electric charge and as such, they consist of both an electric and a magnetic component.

Electromagnetic waves exist with an enormous range of frequencies. This continuous range of frequencies is known as the electromagnetic spectrum. The entire range of the spectrum is often broken into specific regions. The subdividing of the entire spectrum into smaller spectra is done mostly on the basis of how each region of electromagnetic waves interacts with matter. The diagram depicts the electromagnetic spectrum and its various regions. The longer wavelength, lower frequency regions are located on the far left of the spectrum and the shorter wavelength, higher frequency regions are on the far right. Two very narrow regions within the spectrum are the visible light region and the X-ray region.

Visible Spectrum

The first step in understanding color is to examine the nature of light. Light is the visible part of the electromagnetic spectrum, which in turn consists of billions of waves that move through the air like ripples in a pond and surround us at all times. Each wave is a different size, as measured by its wavelength, the length from the wave crest to adjacent crest. Wavelengths are measured in nanometers (nm), or one millionth of a millimeter.

The visible spectrum consists of the wavelengths that can be seen by the human eye. This span lies between about 380 and 720 nanometers, which amounts to a small slice through the middle of the massive electromagnetic spectrum. Although unseen, many of the visible waves on the visible spectrum are used in other ways, from short wavelength x-rays to the broad wavelengths that are picked up from radios and televisions (Image 2).

Eyes have light sensors that are sensitive to the visible spectrum wavelengths. These sensors send signals to the brain, where the perceptual effects of the wavelengths are processed. If these light sensors detect all visible wavelengths at once, the brain detects white light. If the eye detects no wavelengths, there is no light present and the brain perceives black. Black surfaces absorb most of the light that falls on them. This is why a black car is hotter than a white car when parked in the sun. The sunlight is absorbed by the black car and converted into heat energy.

It is known that the eyes and brain respond to the presence of all visible wavelengths or no wavelengths. How the vision system responds to each individual wavelength in the visible range will now be explained. To do this, Sir Isaac Newton's familiar prism experiment will be studied.

Passing a beam of white light through a prism disperses the light to show how the eyes respond to each individual wavelength. This experiment demonstrates that different wavelengths can cause different colors to be seen. This visible spectrum's dominant regions of red, orange, yellow, green, blue, indigo, violet and the rainbow of other colors blending seamlessly in between can be recognized.

When the visual system detects a wavelength around 700 nm, red is seen. When a wavelength around 450–500 nm is detected, blues are seen. These responses are the basis for the billions of colors that the vision system detects every day.

However, all wavelengths are rarely seen at once (pure white light), or just one wavelength at once. The world of color is more complex than that, simply because color is not just part of light, it is light. When color is seen, light has been modified into a new composition of many wavelengths. For example, when a "red" object is seen, light that contains mostly red wavelengths is being detected. This is how all objects get their color, by modifying light. It is possible to see a world full of colorful objects, because each object sends to the eyes a unique composition of wavelengths.

When light waves strike an object, the object surface absorbs some of the spectrums wavelengths, while other parts of the spectrum are reflected back from the object. The modified light that is reflected from the object has an entirely new composition of wavelengths. Different object surfaces contain pigments and dyes in various concentrations and combinations and generate different wavelength combinations or colors, as they are perceived.

Additive Color

The three spectral curves in Images 10, 11 and 12 are for the three additive primaries of red, green and blue. If we were to take all three spectra and add them together, the result would be the spectrum of Image 13, which is white. Of course, adding together the spectra is nothing more than combining or adding the light itself. It is the same as shining three beams of different colored light onto one area. The primary colors of red, green, and blue combine as shown in Image 14.

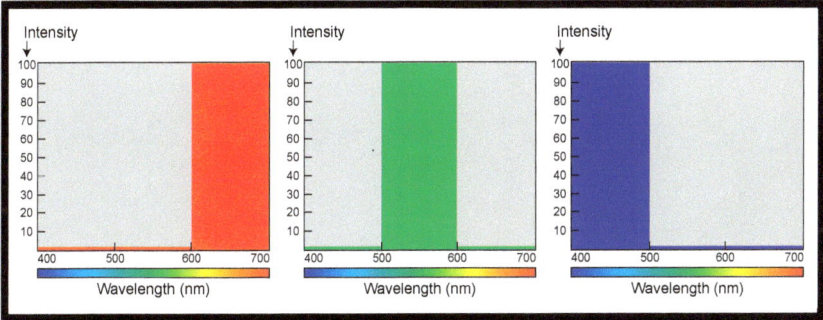

Image 10: Perfect Red   Image 11: Perfect Green   Image 12: Perfect Blue

Before we can understand subtractive color, we must first understand additive color. By adding light, we add color. For the sake of this discussion, let's assume we are sitting in a black room with no light other than a projector. If we turn on the red light in the projector, we see red light. When we turn off the red light, and turn on the blue light, we see blue. When we turn off the blue light and turn on the green light, we see green.

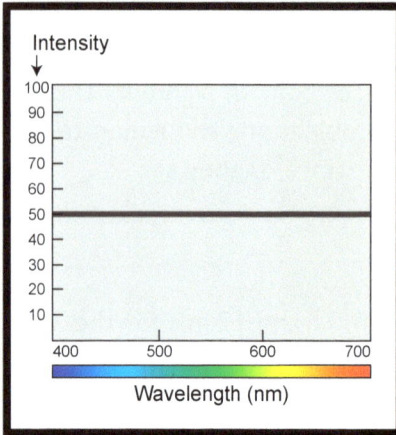

Image 13: Perfect Neutral Spectrum

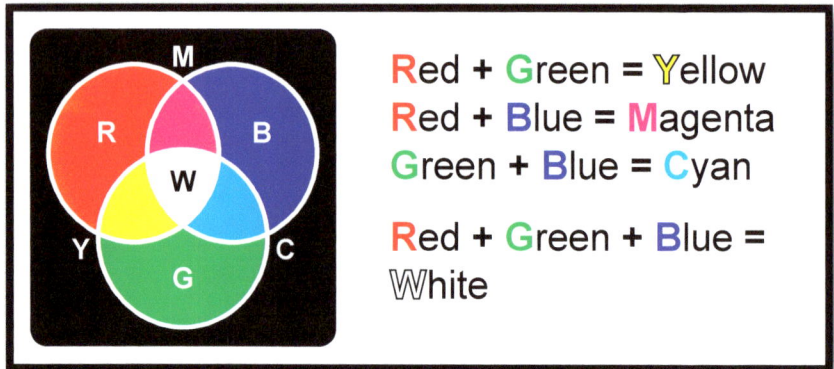

Red + Green = Yellow
Red + Blue = Magenta
Green + Blue = Cyan

Red + Green + Blue = White

Image 14: Red Light + Green Light + Blue Light = White

Overall, this is a very simple concept until we start mixing the color of the light. Looking at Image 14, when we turn on the red and green light, we see yellow. If we turn off the red light and turn on the blue light, leaving the green light on, we see cyan. With the blue and red lights on, we see magenta. With all three lights on in equal strength or intensity, we see white or gray depending upon the intensity of the light. The issue here is that we cannot print with light. Instead, we use printing inks. In all aspects of printing, the inks are formulated to achieve a specific goal or purpose. Regardless of the print principle, the goal is the same.

Subtractive Color

Process printing inks reflect and absorb a portion of the visible color wavelength in order to create the perception of the desired color on the media or substrate. Cyan, magenta, and yellow inks are manufactured to absorb one third of the visible color wavelength and reflect two thirds of the visible color wavelength. Cyan ink is a filter on paper that reflects the blue and green light, and absorbs the red light. Depending upon the purity of the ingredients, and formulation, these inks may absorb and reflect light differently.

Continuing with the theme, magenta ink is a filter on paper that reflects the red and blue light, and absorbs the green light. Lastly, the yellow ink is a filter on paper that reflects the red and green light, and absorbs the blue light. All of the inks printed together would theoretically absorb all of the light, creating black. However, that is seldom the case with printing inks due to the chemical makeup. For this reason, we add black ink to extend the tone range, assist in maintaining gray balance, as well as adding depth in the shadow regions of the image reproduction.

With that said, let's try a practical exercise. We all know that printing yellow and cyan ink on the media creates a green overprint. What we may not understand is how the earlier

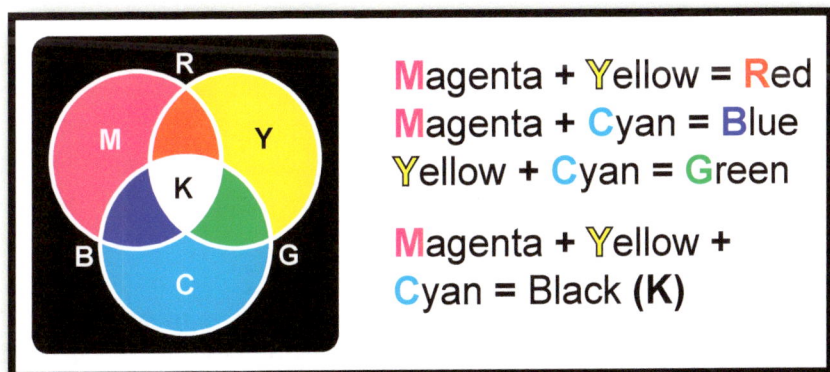

Magenta + Yellow = Red
Magenta + Cyan = Blue
Yellow + Cyan = Green

Magenta + Yellow + Cyan = Black (K)

Image 15: Magenta + Yellow + Cyan = Black

discussion comes into play in this exercise. Assuming a print order of yellow, then cyan, lets break down what happens. The media is formulated to accept ink as well as possessing the ability to reflect the light from the light source (the viewing booth). When the yellow ink is printed on the white paper, the yellow ink acts as a filter on paper. The red and green light is reflected and the blue light is absorbed. Now the cyan ink is printed on top of the yellow ink. The cyan ink reflects the green and blue light, absorbing the red light. If cyan absorbs all of the red light, and yellow absorbs all of the blue light, the only thing left is green, which is what we see.

The same principle holds true for the overprint called red and blue. If magenta is printed on top of yellow, the magenta absorbs the green and the yellow absorbs the blue, leaving the red, which is what we see. For the blue overprint, cyan absorbs the red light, magenta absorbs the green light, leaving the blue light.

Starting with white and taking away one third of the light at a time is utilizing the subtractive primaries of yellow, magenta, and cyan. This is what happens in printing. We start with a white (or at least highly reflective) substrate and the inks we use (cyan, magenta, yellow) each take away roughly one third of the visible spectrum. They, combine as shown in Image 15.

Using this concept of taking away, gives the same result as shown in Image 15. Combining magenta and yellow inks takes away green and blue light, leaving red. Combining magenta and cyan inks, takes away green and red light, leaving blue. Combining yellow and cyan inks, takes away blue and red light, leaving green.

Spectrophotometry
In chemistry, spectrophotometry is the quantitative measurement of the reflection or transmission properties of a material as a function of wavelength. It is more specific than the general term electromagnetic spectroscopy in that spectrophotometry deals

with visible light, near-ultraviolet, and near-infrared, but does not cover time-resolved spectroscopic techniques.

Spectrophotometry involves the use of a spectrophotometer (Image 16). A spectrophotometer is a photometer that can measure intensity as a function of the light source wavelength. Important features of spectrophotometers are spectral bandwidth and linear range of absorption or reflectance measurement.

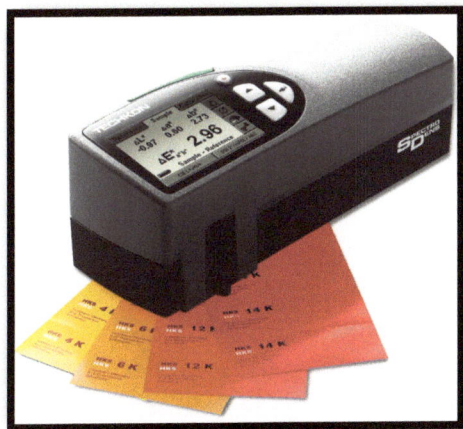

Image 16: Techkon Spectrodensitometer

The Spectrophotometer

A spectrophotometer is used to measure the entire visible spectrum of a sample. The key part of the spectrophotometer is an optics head that contains a light source in a fixed geometry, an element like a prism which breaks up the light into its discrete wavelengths, and a detector of the dispersed light. The spectrophotometer can either display the measurement data spectrum, or it can send the measurement data spectrum to a computer. Measurement data can be CIEL*a*b*, L*C*h°, or spectral data, depending upon the device and its capabilities.

The ability of the spectrophotometer to do calculations enables it to also function as a densitometer. In a densitometer, the light

Color Densitometer	
Filter	Measures
RED	CYAN
GREEN	MAGENTA
BLUE	YELLOW
VIS	BLACK

Table 1: Color Densitometer

is filtered as shown below in Table 1. This is nothing more than a modification of the light source. If the spectrum of the filter is known, all densitometric values can be calculated. The spectra of the measurements have been defined and are known as filters, such as Status T, Status E, and the like.

The Status T filter is predominantly used in the United States. Using this standard, all the metrics used in densitometry can be calculated by a spectrophotometer. Physically, a reflection spectrophotometer looks very similar to the reflection densitometer illustrated in Image 17. L*a*b* or L*C*h° values are a combination of the object and source spectra, taking into account the response of the standard observer. The optics head delivers the object spectrum and the standard observer is well defined and fixed. The effect of different light sources, such as D50 and D65, can be calculated, and the spectrophotometer can display the resulting L*a*b* or L*C*h° values under these different sources.

The spectrophotometer can make absolute measurements or measurements relative to the substrate. The spectrophotometer can be used to quantify metamerism. As an example, Image 18 shows the spectra of the two colors in the RHEM light indicator. Note that the two spectra cross at several points, a condition required for two colors to be metameric.

Illuminated with D50 light, the colors match to a CMC ΔE of less than 1. Illuminated with "A" light, the colors match to a CMC ΔE of 2.86, which is clearly visible.

A common measure of metamerism is called the metamerism index (MI), which can also be calculated. In this case, its value is 3.6. The metamerism index measures the difference between the colors under different light conditions. A low value does not mean the colors are the same, only that the visual difference is the same under both light conditions.

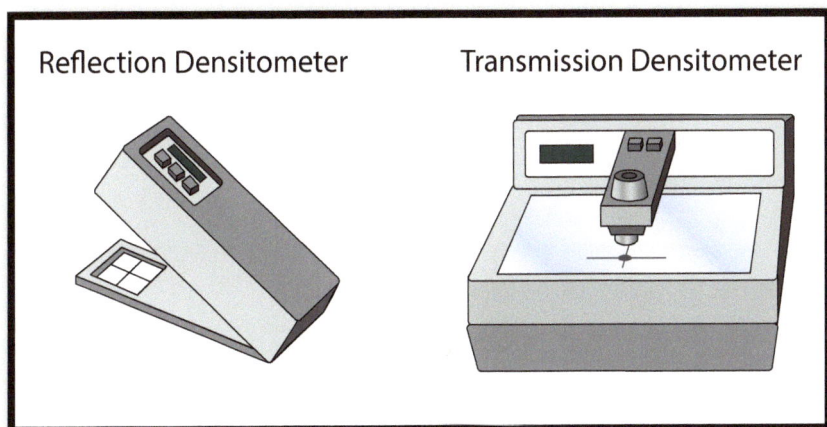

Reflection Densitometer   Transmission Densitometer

Image 17: Reflection Densitometer and Transmission Densitometer

Spectrophotometers work in a similar way to spectral-based colorimeters. They split the visible spectrum into very small segments using either narrow-band filters or a diffraction grating. All spectrophotometers can output the same data as colorimeters, however, the spectrophotometer is a more sophisticated instrument and able to output the information as a spectral curve. This curve is derived from taking the percentage reflectance at each wavelength measured and plotting it on a graph. Once each point has been plotted, the dots are connected to produce a curve that is unique to each pigment color measured. These curves can be used like a fingerprint to identify the pigments that make up an ink.

The spectrophotometer is the ideal instrument to use when mixing inks. The instrument can save a great deal of time spent on hit-and-miss ink mixing. Its use will improve the batch-to-batch consistency of ink, along with ensuring consistency between different ink department individuals.

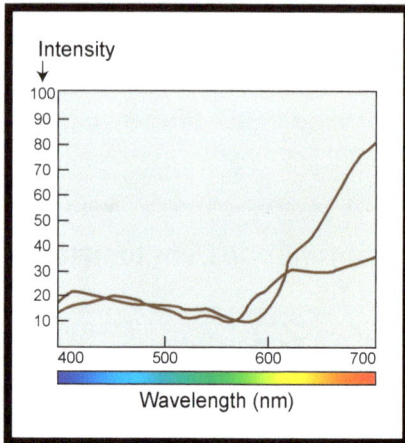

Image 18: Wavelength Intensity

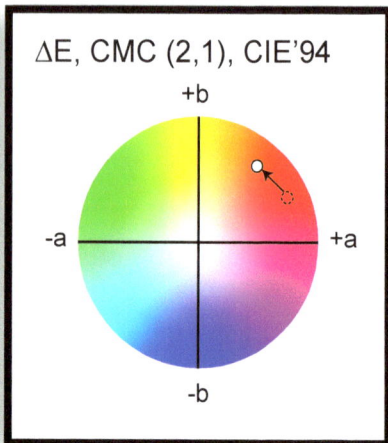

Image 19: Delta E (ΔE)

A spectrophotometer is used to measure the entire visible spectrum of a sample. The real color curves presented elsewhere in this booklet, were all taken with a spectrophotometer. The key part of the spectrophotometer is an optics head that contains a light source in a fixed geometry, an element like a prism that breaks up the light into its discrete wavelengths, and a detector of the dispersed light. The spectrophotometer can either display the spectrum, or it can send the spectrum to a computer.

COLOR COMPARISON

Once a color is described in terms of a point in space, the concept of a color difference follows naturally. It is the geometric distance between two colors (Image 19) and is called delta E (ΔE).

As a measure of the difference between two colors, ΔE serves as a specification of color tolerance. That is, two colors match if their difference is less than a certain value of ΔE.

Unfortunately, specifying an acceptable ΔE value is not a simple matter. Ideally, the same ΔE would mean the same perceived color difference throughout color space. Experience shows that this is not the case. A small ΔE in a neutral gray would be more apparent than the same ΔE in a saturated dark red.

To overcome this deficiency, weighting factors are introduced into the ΔE calculation. Currently, the CMC weighting calculation has widespread acceptance. With some modification, this has been adopted by the CIE as CIE'94. When quoting ΔE values or tolerances, it is essential to know which calculation is being used. Otherwise, the numbers will be different. Typically, reference is made to ΔE, CMC or CIE'94 tolerance or color difference. To complicate matters even further, there are additional adjustment parameters used in the CMC and CIE'94 calculations. The usual values for these are 2 and 1 and the CMC color difference may be quoted as CMC (2,1).

Visual
To compare, communicate and store color data, it is necessary to adopt a measurement system. The human visual system is the most discriminating when comparing colors, but it is neither able to assign numbers to colors, nor remember them accurately. That is why some sort of a numerical measurement standard and an organized method of communicating color is needed.

If an object is viewed using other than a white-light source, the perceived color will take on the hue of the illumination. While sunlight is the most natural way to view objects, it is not an ideal light source to judge the color of objects; it is simply too variable.

Artificial light sources are available and may be controlled and specified to simulate average, natural daylight, and incandescent and fluorescent lamps.

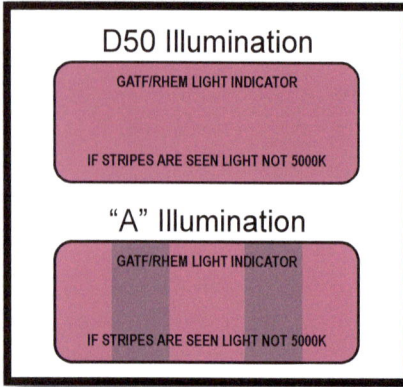

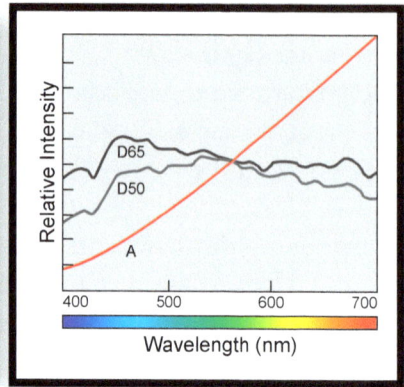

Image 20: RHEM Light Indicator   Image 21: Spectra of Standard CIE

A scale of color temperature, expressed as degrees Kelvin (°K), is used to quantify light sources. Various artificial light sources have color temperatures that range from about 4,000° K to 6,800° K. The D50 CIE Standard Illuminant has a color temperature of 5,000° K, representing average natural daylight, and is the color temperature most widely used in graphic arts viewing booths. Image 21 shows three light sources along with their spectral curves. The spectral curve shows the amount of light of the source throughout the visible range of wavelengths, which is roughly from 400 to 700 nanometers.

Spectral Reflectance
Spectral Reflectance is a curve that illustrates the reflectance of light from a surface, such as paper, wavelength-by-wavelength throughout the visible spectrum, as a means of determining the color of that surface. Spectral reflectance curves are used to determine paper color, brightness, and whiteness, important optical properties that dramatically affect the quality of material printed on the paper surface.

Metamerism
Every color has a unique point in CIEL*a*b* color space, its own set of L*a*b or L*C*h° values. What is not unique is the

combination of spectral curves of the source and object, which can produce that color.

This leads to the common phenomenon called metamerism. It means two colors are a match under one illumination, but not under a second illumination. Visually, the test for metamerism simply means looking at the sample under different illumination sources. Many light booths provide multiple sources for this purpose. If no light booth is available, the sample can be viewed near an open window to approximate D50 and then under a standard tungsten filament light bulb (illuminant A). A quick, inexpensive and less rigorous method to determine if a light source is the standard D50 is to use the RHEM light indicator. Available from the Printing Industries of America is an illumination test target consisting of alternating patches of two colors that match under D50, but do not match under different illumination such as "A" or standard fluorescent lights. Image 20 illustrates a simulation of the indicator (actual appearance will be different). Simple visual examination reveals if the illumination is D50 (or at least close to it). Similar illumination test targets are available from other vendors.

Light Sources

Different light sources emit light, which have different spectra. Table 2 lists CIE standard light sources or illuminants, while Image 21 shows the spectra of "A", D50 and D65. "A" is for a tungsten filament bulb (i.e., an ordinary light bulb) at a color temperature of 2,850° K. D50 and D65 represent light at color temperatures of 5,000° K and 6,500° K respectively.

Degrees Kelvin is a temperature scale much like degrees Fahrenheit (° F) and degrees Celsius (° C). The different temperatures mean that a well-defined material heated to that temperature would emit light of a given spectral composition. This is called black body radiation or the black body curve. In the

CIE Standard Illuminants	
Illuminant	Description
A	Incandescent lighting at a color temperature of about 2,850° K
B	Direct sunlight at about 4,874° K
C	Tungsten illumination simulating daylight at about 6,744° K
D50	Graphic arts standard viewing conditions at about 5,000° K
D65	Used by textile, paint and ink industries, about 6,500° K
F2	Cool-white fluorescent lamp at about 4,200° K
F7	Broad-band daylight fluorescent lamp at about 6,500° K
F11	Narrow-band white fluorescent lamp at about 4,000° K

Table 2: CIE Standard Illuminants

graphic arts, D50 or 5,000° K light is standard for making color evaluations. The light sources themselves are special types of fluorescent light bulbs.

Note: As with any specification, nothing is ever exact, there is always a tolerance. For a D50 source, one measure of this tolerance is called the color-rendering index (CRI). The higher this number, the more closely the source matches D50, with 100 being a perfect match. For color evaluation in a light booth, a rendering index higher than 90 should be used. The point to be made here is that not all D50 light tubes are created equal.

Tolerancing
Color tolerancing is used to determine if a "sample" color, compared to a "reference" color, is acceptable. The accuracy of the color match is typically expressed as Delta E (ΔE or DE). The ΔE value represents the overall color difference and is derived from one of several equations. All color difference equations

compare the location of a "reference" color to the "sample" color in CIEL*a*b* color space; however, CIEL*a*b* color space is not uniform. For example, shifts in hue are typically more easily perceived than shifts in lightness or chroma. Color difference equations are either weighted or unweighted.

Unweighted color difference equations, such as CIE 1976 (ΔEab), weight hue, chroma and lightness equally. Equal weighting does not correspond well with human perception of color differences; therefore, a sample may visually match the reference color but produce an unacceptable ΔE value or vice versa.

Weighted color difference equations, such as DECMC, DECIE94, and DE2000, achieve better agreement with human perception by weighting hue, chroma, and lightness differently. The formula used to weight the three-color axis is slightly different for each weighted color difference equation.

The CMC tolerancing method includes three weighting factors set by the user, expressed as (l:c:cf), where: l=lightness, c=chroma, and cf=commercial factor, or ΔE. The CMC ratio l:c (lightness: chroma) determines the shape of the ellipsoid, which is typically set at 2:1 for most applications; however, an l:c ratio of 1:1 is becoming more common. The cf (commercial factor) determines the overall size of the ellipsoid and the threshold, or tolerance, of acceptable color difference. The cf determines the ΔE limit, for example, if the cf = 1.5, then the acceptable ΔE = 1.5 as well.

FIRST recommends using one of the weighted color difference equations and, where possible, the latest applicable equation. Regardless of the color tolerancing equation used, it is critical to communicate both the equation and any weighting factors used to all parties receiving color data.

All parties measuring color must use the same tolerancing equation in order to have meaningful discussions of color

Comparison of Color Difference Equations					
Parameter	CIE76	CMC		CIE94	CIE2000
FIRST Recommended?	No	Accepted		Accepted	Preferred
Year Developed	1976	1984		1994	2000
Equation Developed By:	The International Commission on Illumination	The UK Society of Dyers & Colorist Color Measurement Committee		The International Commission on Illumination	The International Commission on Illumination
Pass/Fail Ellipsoid around the Standard with a set DE?		Yes		Yes	No
Is the DE the same if the "Standard" and "Sample" are reversed?		No		No	Yes
Weighted?	No	Yes		Yes	Yes
Typical Weighting Factors (for print applications)	N/A	(2:1)	(1:1)	K1 = 0.045 & K2 = 0.015	(1:1:1)
Example DE Comparison: "Standard" L= 45.25, a= -40.35, b= -35.75 "Sample" L= 43.75, a= -39.50, b= -37.50	2.46	1.22	1.75	1.83	1.65

Table 3: *FIRST* Comparison of Color Differences Equations

differences. Indicate the color tolerancing equation used on all proofs or print samples containing color difference (ΔE) values. As instrumentation is purchased, *FIRST* recommends buying equipment that supports the latest methods in color tolerancing and has a robust upgrade path.